ELL Strategies
for Science

Reviewers

Dr. Laurie Weaver
Assistant Professor
Bilingual, ESL and Multicultural Education
University of Houston-Clear Lake
Clear Lake, Texas

John B. Martinez
Coordinator for Reading and Language Arts
McAllen Independent School District
McAllen, Texas

D1361401

Glencoe
McGraw-Hill

New York, New York   Columbus, Ohio   Woodland Hills, California   Peoria, Illinois

Glencoe/McGraw-Hill

A Division of The McGraw·Hill Companies

Send all inquiries to:
Glencoe/McGraw-Hill
8787 Orion Place
Columbus, OH 43240

ISBN 0-07-829661-7
Printed in the United States of America
6 7 8 9 10 024 06 05

Table of Contents

To the Teacher

English language learners (ELLs) are students whose primary language is not English. More than 6 million students in the United States speak a language other than English in their homes. The level of English proficiency among these students varies greatly—from students who have recently arrived in the United States and are hearing English for the first time, to students who have a good command of the English language. Just as the level of proficiency varies greatly among English language learners, so do the variety of primary languages and cultures. Many students have emigrated from Spanish-speaking countries, as well as Africa, Asia, and the Middle East.

The rapid increase in culturally and linguistically diverse students presents certain challenges to teachers. While some school districts may offer sheltered instruction for English language learners, the more typical model includes an hour or two of separate English language instruction per day with all students mainstreamed into English-speaking classes for science, mathematics, and social studies.

Many teachers of core subjects have limited experience in teaching English as a second language (ESL). While helping students learn a new language, they also must focus on teaching a core subject. This can be a difficult challenge when coupled with demands made by state-mandated curricula and tests.

Glencoe's *ELL Strategies for Science* is a guide for science teachers who have English language learners in their classrooms. It provides practical tools and suggests modifications that can help students master scientific concepts while developing their English language skills.

ELL Strategies for Science focuses on methods for successful inclusion of English language learners in the science classroom. These strategies rely not only on teacher intervention, but also on student intervention to create ownership of the learning process.

For those teachers just being introduced to teaching English as a second language, *ELL Strategies for Science* provides an overview of how students learn a language. The overall approach is compatible with research that shows that English language learners are most successful when using a hands-on, inquiry approach with clearly articulated learning strategies.

The teaching models in *ELL Strategies for Science* are structured around major science concepts, focusing on subject depth rather than breadth. The overall objectives incorporate *National Science Education Standards* and can be easily customized to match state-specific standards. The learning strategies are those identified by educators as being the most useful in teaching scientific content to English language learners. Methods include many lower-order thinking skills, such as fact recall and identifying and defining vocabulary, as well as higher-order critical thinking skills, such as comparison, classification, and prediction.

Many ESL teaching strategies are presented in *ELL Strategies for Science*, including those suggested in the Cognitive Academic Language Learning Approach (CALLA). The CALLA method asserts that content, not language, should drive the curriculum. It integrates content area instruction with academic language development and specific learning strategies.

In addition to the information presented in *ELL Strategies for Science*, you may find other Glencoe science products helpful in teaching English language learners. See the resource list on the next page.

Glencoe Science Resources
Teacher Support
- Performance Assessment in the Science Classroom
- Home and Community Involvement
- Cooperative Learning in the Science Classroom
- Dinah Zike's Teaching Science with Foldables
- Content Outlines for Teaching
- Spanish Resources
- ExamView
- PuzzleMaker

Student Resources and Activities
- Reading and Writing Skill Activities
- Cultural Diversity
- Transparency Activity Masters
- Directed Reading for Content Mastery
- Directed Reading for Content Mastery in Spanish
- Reinforcement
- Enrichment
- Note-taking Worksheets
- Guided Reading Audio Program
- Interactive CD-ROM with Presentation Builder
- Video Labs
- MindJogger VideoQuizzes

English Language Learners in the Science Classroom

A machine reduces the effort force needed to counter a resistance force.

Imagine walking into a classroom and seeing this sentence written on the chalkboard. If you are an English speaker with some science background, you would probably recognize its general meaning. Now, consider how students just learning English might read this sentence. Depending on their proficiency, they might recognize the subject as "machine" and the verb as "reduces." Other words in the sentence would be more difficult. Consider the use of "effort" and "resistance." Both words are commonly used as nouns. In this sentence, both are used as adjectives to describe "force"—a word that can be a verb or a noun. "Counter" is used as a verb in this context, but dictionaries list at least five other definitions. This is just one example of an obstacle English language learners face in the science classroom.

As science classrooms have become more ethnically and culturally diverse, teachers are scrambling to find the best way to convey scientific concepts, while at the same time teaching a new language. In the last few years, researchers have been working to come up with effective ways of integrating classroom content and language skills. This handbook will provide some of the results of that research along with tools and models to help you with this challenge.

Who Are English Language Learners?

English language learners, abbreviated as ELLs, are students whose primary language is not English. These students are learning English as their second language. Some have attended school in another country and are at the same grade level as other students their age. Others may have had to interrupt their education for one reason or another and may require additional assistance through the use of intervention strategies. Although ELLs are generally willing to participate in second language acquisition classroom activities, many often revert to speaking their primary language in school settings and in their home environment, thus limiting their exposure and practice with the new language.

Some ELLs have come from countries that have cultures similar to ours and will have to make minor social and cultural adjustments to adapt to their new surroundings. Others, from countries where the culture is characteristically unique, will have to make major adjustments to adapt to their lives in a new city, a new neighborhood, and a new school.

Like the rest of the students in your classroom, ELLs have as many differences as they do similarities. What works for one student will not work for all students. Just like English-speaking students, ELLs have multiple learning styles. Some achieve their goals quickly and independently. Others take more time and more teacher assistance to accomplish their goals.

As a teacher, you will need to evaluate the level of proficiency of your ELLs, just as you would with your English-speaking students. The important thing to remember about English language learners is that they are trying to master a new language at the same time they are mastering academic content.

The Teachers of English to Speakers of Other Languages (TESOL) has created standards to help ELLs improve their educational outcomes. The standards set three goals:

Goal 1: To use English to communicate in social settings

Standard 1: Students will use English to participate in social interaction.

Standard 2: Students will interact in, through, and with spoken and written English for personal expression and enjoyment.

Standard 3: Students will use learning strategies to extend their communicative competence.

Goal 2: To use English to achieve academically in all content areas

Standard 1: Students will use English to interact in the classroom.

Standard 2: Students will use English to obtain, process, construct, and provide subject matter information in spoken and written form.

Standard 3: Students will use appropriate learning strategies to construct and apply academic knowledge.

Goal 3: To use English in socially and culturally appropriate ways

Standard 1: Students will choose appropriate language variety, register, and genre according to audience, purpose and setting.

Standard 2: Students will use non-verbal communication appropriate to audience, purpose, and setting.

Standard 3: Students will use appropriate learning strategies to extend their sociolinguistic and sociocultural competence.

Evaluating Language Proficiency

To work effectively with ELLs, teachers must know their level of language proficiency. Obviously, the needs of beginning English learners will differ from those of advanced learners. By recognizing and understanding each student's proficiency level, teachers will be armed with the information they need to modify their classroom instruction.

Some sources break down proficiency into four main categories—novice, beginning, intermediate, and advanced. The term "novice" applies to someone who has been in the United States for a short time (six months or less) and has a minimal ability to communicate and understand English. At the other end of the scale is the advanced learner, whose reading, writing, listening, and speaking skills approximate that of a native English speaker. *ELL Strategies for Science* focuses mainly on beginning and intermediate level students.

Beginning Learners

Beginning learners usually have been in the United States two years or less. Their language skills are in the beginning stages. Beginning learners may be able to understand social language if it is spoken slowly and repeated several times. However, they often are hesitant to speak because of their limited vocabulary and pronunciation problems. Beginning learners need structure in the classroom and support from both teacher and students.

Intermediate Learners

Intermediate learners have been in the United States between two to five years. They have well-developed social language skills but lack the academic language skills required in a content classroom. The fact that they speak so well conversationally often leads teachers to believe that they are at a higher level than they really are. Intermediate learners are capable of comprehending, reading, and writing in English. They can work well independently but still benefit from teacher support and support from students, especially in cooperative learning groups. With intermediate learners, it is important to define major content areas and overtly suggest learning strategies to help them master content and higher-order thinking skills.

The following checklists were designed to help teachers modify their instructional and assessment approaches for ELLs who are at the beginning and intermediate levels of language proficiency. They were developed by ESCORT, a national resource center at the State University of New York, as part of a kit designed to help secondary teachers of migrant English language learners. The project was funded by the U.S. Department of Education.

Beginning Learners

(Up to two years in an English-speaking classroom with ESL/bilingual support)

Instructional Modifications

All Students

___ Use visuals/"hands-on" manipulatives

___ Use gestures to convey meaning non-verbally

___ Provide concrete "real" examples and experiences

___ Build on the known (e.g., make connections with students' culture, experiences, interests, and skills)

___ Simplify vocabulary/change slang and idioms to simpler language

___ Highlight/review/repeat key points and vocabulary frequently

___ Establish consistent classroom routines/list steps for completing assignments

___ Use yes/no, either/or, and why/how questions (Allow wait time for response)

___ Check for comprehension on a regular basis ("Do you understand?" is not detailed enough)

___ Create story and semantic maps

___ Use "Language Experience Approach"

___ Plan ways for ELLs to participate in class and in cooperative learning groups

___ Make outlines/use graphic organizers

___ Use audiotapes to reinforce learning

___ Use simplified books/texts that cover content-area concepts

___ Translate key concepts into a student's native language

Students with adequate literacy in their native language (if bilingual person is available)

___ Use textbooks/books in the native language that cover key concepts being taught

___ Encourage student to use a bilingual dictionary as a learning tool

___ Have student write essays/journal entries in the native language

Assessment Modifications
All Students
___ Have student point to the picture of a correct answer (limit choices)

___ Have student circle a correct answer (limit choices)

___ Instruct student to draw a picture illustrating a concept

___ Reduce choices on multiple-choice tests

___ Instruct student to match items

___ Have student complete fill-in-the-blank exercises with the word list provided

___ Give open-book tests

___ Ask student to retell/restate (orally and in writing)

___ Instruct student to define/explain/summarize orally in English or the native language

___ Have student compare and contrast (orally and in writing)

___ Use cloze procedure with outlines, charts, timelines, etc. (This procedure is a technique in which words or phrases are deleted from a passage. Students must fill them in as they read.)

Students with adequate literacy in their native language (if bilingual person is available)
___ Instruct student to write what he or she has learned in the native language

Intermediate Learners
(Up to five years in an English-speaking classroom with ESL/bilingual support)
Assessment Modifications
All Students
___ Instruct student to explain how an answer was achieved (orally and in writing)

___ Have student complete fill-in-the-blank exercises

___ Ask student to retell/restate (orally and in writing)

___ Instruct student to define/explain/summarize (orally and in writing)

___ Have student compare and contrast (orally and in writing)

___ Use cloze procedure with outlines, charts, time lines, etc.

___ Have student analyze and explain data (orally and in writing)

___ Instruct student to express opinions and judgements (orally and in writing)

___ Have student write essays

Finding Out What Works

At one time, English language learners were confined to special classes where only language skills were taught. This practice assumed that students had to be proficient in English before learning any content. This not only isolated them from other students but also delayed their instruction in content areas. Current research shows that the best approach is an integrated approach—using content as a way of teaching English language skills. Researchers have found that many students are more motivated when they are learning about a subject than when they are learning language alone. Science is particularly motivating because of its hands-on approach and students' inherent interest in scientific topics.

Learning a New Language

What's it like to learn a new language? While research has shown that it takes most English language learners about two years to learn enough English to communicate in social situations, it takes between five to seven years for students to master the academic language necessary to succeed in a classroom.

Social language is the language of everyday living. It is easier to learn than academic language because people are immersed in it in their neighborhoods, in stores, on television, and even on the Internet. In addition, there are many non-verbal cues that aid English language learners in a social situations, such as facial expressions.

Academic language is the language students need to help them understand and process a subject such as science. Mastering academic language is more difficult than mastering social language because it includes more than just terminology. In a subject such as science, students also must learn to think scientifically. This involves the use of lower-order thinking skills, such as recalling facts, as well as developing higher-level critical thinking skills that will help them succeed in the science classroom and laboratory. Observation, creating hypotheses, designing experiments, drawing conclusions, making predictions, and presenting information in a variety of scientific formats are all part of the scientific learning package.

Some Misconceptions about ELLs

One misconception about English language learners is that if they have a good command of social language they will do well in the content classroom. This is not always the case and points to the necessity of carefully assessing the language proficiency of ELLs.

It also is important to realize that even if students are unable to articulate their knowledge of content in a particular area, you cannot assume that they do not possess that knowledge. Many ELLs find it difficult to speak in a new language in front of their peers in the classroom. Until their confidence grows, it is helpful to let them express their knowledge in other ways. For example, ask them to draw a picture or label a diagram to illustrate a particular concept. If they are more confident speaking in small groups, create an activity that focuses on that strength.

Another misconception is that if the content is watered down students will benefit by being able to understand more. There are problems with this assumption. One is that ELLs may have already achieved proficiency of the content in their native language. Instead of reducing the level of the material, the best solution is to help students retrieve and convey the knowledge they have already acquired. A second problem with diluting content is that it will isolate ELLs from the rest of the students in your classroom. Rather than providing lower-level content, a teacher's time is better spent modifying the content to make it more accessible to ELL students. Strategies will be discussed in more detail later in the manual.

How Do I Create the Best Environment for Learning?

Before discussing specific approaches to teaching science to English language learners, there are some general guidelines that will help you to create a good environment for learning in your science classroom. In general, flexibility and creativity will go a long way when creating a positive and practical classroom environment.

One Size Does Not Fit All

There is no magic formula for determining the *one* best learning strategy or activity to use in teaching a particular scientific concept. The best results most often come from a combination of approaches and strategies. Whenever possible, focus on multisensory approaches in a variety of media. The variety will not only assist your ELLs but also prove to be a benefit for your English-speaking students.

Remember, what works for one student may not work for another. Be flexible and be willing to try new approaches. Continually develop and add to your repertoire as you discover approaches that work.

Learn About Your Students

Just as knowledge of your students' proficiency will help you understand their abilities and learning styles, knowledge of their background and culture also will help in making your instruction more effective. Keep an information card on each student. You may want to sit down with each student and fill out information cards together. Ask an ESL instructor or another student to serve as a translator if needed.

Student Information Card

Name: _____

Country of origin: _____

Date student moved to the United States: _____

First language: _____

Language spoken at home: _____

Last grade completed in home country: _____

Name(s) of possible translator: _____

Language strengths: _____

Lanuage weaknesses: _____

Goals: _____

_____

Focus on Depth Rather than Breadth

Concentrate your efforts and your students' efforts on learning major scientific concepts rather than trying to learn a little about all scientific concepts. A standards-based curriculum can help you narrow down the most important themes.

Speak Clearly and Precisely

Critique your classroom speech habits to pinpoint areas that might be confusing to ELLs. It is not necessary to water down scientific material or vocabulary, but material does need to be presented clearly and precisely for the best results.

a) Use active voice as much as possible. For example, instead of "Ozone is produced by the burning of fossil fuels, " say, "Burning fossil fuels produces ozone."

b) Be specific about the terms that a student needs to learn when introducing a new science concept. For example, when introducing the rock cycle, reinforce to students that they will need to learn the terms *rock cycle, igneous, metamorphic, sedimentary,* and so forth.

c) If students are having trouble with comprehension, use synonyms or paraphrase to clarify a major concept.

d) Indicate sequence clearly by adding tag words such as *first, then,* or *finally.*

e) Use a variety of contextual and visual examples in presenting lessons. Graphs, charts, diagrams, as well as flowcharts, time lines, and other graphic organizers can often convey information that the spoken word cannot.

f) Serve as a model for your ELLs and try to get your English-language students to do the same. Use language the way that you would like your students to use language.

Clearly Define Expectations

When making assignments, be specific about what students are expected to know, to do, and to submit for a passing grade. It is particularly helpful to develop a set of guidelines or a scoring rubric. Use samples to model the expected outcome whenever possible. You may want to use an overhead projector or the chalkboard to show a sample of a lab report, diagram, or other assignment. Go through it line by line or element by element and model what information should appear.

Foster Independence

Empower ELLs by making them self-reliant. Let them make choices about the learning strategies that work best for them. Use scaffolding to provide support when teaching a new concept, and then gradually pull back as a student becomes more self-sufficient. By providing students with options of how they can learn, it allows them to choose what works best for them. This approach also serves as a motivator and reinforces confidence.

Ask Questions

Asking questions can help you gauge a student's linguistic ability and assess prior knowledge and comprehension. "Yes" and "no" questions are sometimes less confusing for beginning ELLs. As students become more proficient, questions should prompt both lower-order and higher-order thinking.

Encourage Interaction

Whenever possible, group students in pairs, small groups, or cooperative learning situations to encourage verbal interaction and allow ELLs to use other students as models for their communication. Use role-playing situations to reinforce concepts as well as to take advantage of students' more developed social language skills. Pair ELLs with English speakers to encourage collaboration and attainment of common goals.

Make Your Classroom Environment Multicultural

When discussing scientific achievements, focus on scientists and people from around the world. If possible, draw examples from the cultures of your ELLs. Encourage all students to compare and discuss different cultural approaches.

Infuse Your Lessons with Real-World Examples

Refer to everyday activities—throwing a ball, turning on a faucet, what students had for breakfast—to convey science concepts. Rely heavily on authentic assessment to help students make the bridge between their languages and English.

Build Confidence

Model correct language but don't constantly correct a student. If a student answers correctly but uses improper language, provide a response that models the correct language. Incorrect verb choice or noun-verb agreement does not mean that the student does not understand the concept. Remember that speaking in a new language can be very difficult for ELLs. Provide other forms of communication opportunities when possible.

To build confidence and also help in the learning process, expose ELLs to native speakers as much as possible. Rely on cooperative learning activities, role-playing, and other interactions to provide those opportunities.

Acknowledge and respect the cultures of ELLs and encourage all students to do the same. Whenever possible, provide opportunities for your ELLs to succeed.

Provide Classroom Resources

Long reading assignments and classroom lectures can be daunting to ELLs. Textbooks are a valuable resource in and out of the classroom. However, the sheer volume of material may be difficult for ELLs to comprehend. Set aside time to describe the organization of your science textbook. Point out the lists of terms at the beginning of each chapter. Show students where end of chapter and end of unit review questions can be found. Point out the location and use of the glossary and index at the end of the book. Encourage students to use the organization of their textbooks as a model to create a study outline for major science concepts.

Try to provide a variety of resources in your classroom. This is helpful to ELLs but also is an excellent way to supplement and enrich the learning experience for your English-speaking students.

> **Helpful classroom resources include:**
> - a science listening center where students can listen to recordings of class discussions and lessons
> - dictionaries in a variety of languages
> - scientific primary sources
> - samples of completed lab reports, tests, and homework assignments to serve as guides
> - a variety of posters, diagrams, graphs, and graphic organizers that present major scientific concepts
> - pictures or samples of lab equipment labeled in English

Working within Your State or District's System

The structure in place for teaching English as a second language (ESL) varies from state to state and district to district depending on the ethnic population of the area. Some schools have separate classes for ELLs. Others heterogeneously group ELLs with native-English speakers. Some schools have both types of classroom structures. Whatever program is in place in your state, you will need to follow its specific curriculum and assessment guidelines in tailoring your classroom approach.

The National Research Council, in the publication *National Science Education Standards,* has defined scientific literacy as "the knowledge and understanding of scientific concepts and processes required for personal decision making, participation in civic and cultural affairs, and economic productivity."

According to these standards, "Scientific literacy means that a person can ask, find, or determine answers to questions derived from curiosity about everyday experiences. It means that a person has the ability to describe, explain, and predict natural phenomena. Scientific literacy entails being able to read with understanding articles about science in various forms of media and to engage in social conversations about the validity of the conclusions. Scientific literacy implies that a person can identify scientific issues underlying national and local decisions and express positions that are scientifically and technologically informed. A literate citizen should be able to evaluate the quality of scientific information on the basis of its source and the methods used to generate it."

To meet these expectations, for both ELLs and English-speaking students, the nation's science classrooms have had to make some changes. Instead of focusing on a set of specific facts and procedures, instruction has broadened to incorporate methods that encourage scientific discovery through a hands-on approach. Observing, formulating questions, hypothesizing, designing experiments, making predictions, and reporting information in a variety of ways are now central to science instruction.

What is interesting to note is that this same type of approach has also been found to be effective for English language learners. A hands-on technique that incorporates a wide range of discovery strategies provides a means for English language learners to use and reinforce their written and oral language in authentic ways. This match provides a helpful bridge for science teachers who are welcoming ELLs into their classrooms.

Content-Based Language Instruction

In This Section

In recent years, educational specialists have done significant research to determine the most effective way to teach content to English language learners. Many models focus on teaching content and language simultaneously.

One such model is the Cognitive Academic Language Learning Approach (CALLA). It was developed by Anna Uhl Chamot and J. Michael O'Malley. In the CALLA model, *content* becomes the means by which language skills are learned, not vice versa. The model assumes that students will master academic language as they need it to understand and process content. Academic language becomes the means by which students communicate major concepts and processes particular to a subject area.

Much of the early research for the CALLA approach involved interviews with actual ELLs. Many were high-achievers who had developed specific learning strategies to help them learn material in different content areas. CALLA suggests that ELLs need to be prepared with an arsenal of learning strategies to be successful in a content-based classroom.

The CALLA model focuses on three goals:

1. selecting major topics from a content area
2. developing the academic language skills needed to understand and process the material from that content area
3. determining specific learning strategies that can aid in the learning of that content and language

1 • Selecting Content

When teaching content to ELLs, it is important to narrow the range of the content. Students are most successful if they focus on only the most important concepts—concepts that will follow them throughout their science education. The complexity of these concepts increases as students progress from grade to grade, but the basic concepts remain the same.

Fortunately, because of recent reforms in science education, content has received a great deal of attention. Much thought has gone into the specific themes and topics that need to be taught in the science classroom. National, state, and district standards now define the fundamental content areas on which science teachers need to focus. By aligning science content with a particular state or district's curriculum standards, science teachers can be assured of incorporating the most important content areas into their curriculum.

As a general guide, the *National Science Education Standards* state that content is fundamental if it

- represents a central event or phenomenon in the natural world.
- represents a central scientific idea and organizing principle.
- has rich explanatory power.
- guides fruitful investigations.
- applies to situations and contexts common to everyday experiences.
- can be linked to meaningful learning experiences.
- is developmentally appropriate for students at the grade level specified.

2 • Developing Academic Language Skills

Academic language is defined as the language functions or skills that a student needs to understand academic content. It includes the lower- and higher-order thinking skills needed to complete activities in that content area.

The academic language of science is unique and increases in complexity as students move from one grade to the next. In addition to learning scientific terminology, ELLs must learn how to think scientifically using scientific reasoning and critical thinking skills. The hands-on approach to science also requires that they learn to perform scientific processes and know how to communicate scientific theories, results, and predictions in a variety of written and oral forms.

3 • Using Explicit Learning Strategies

To be successful, ELLs have to be armed with specific learning strategies. Learning strategies are activities or thoughts that help students to learn and must be customized to fit the particular activity or task.

By choosing from a menu of specific strategies before they immerse themselves in a science task, research shows that ELLs can increase their likelihood of success. Some students are able to identify these learning strategies on their own. Others will need assistance. In time, students will begin to understand which strategies work best for them without prompting.

A suggested format for teaching learning strategies is to first name the strategy and then discuss when and why to use the strategy. Next, the teacher should demonstrate the use of the strategy. Following the demonstration, students should practice using the strategy in small groups. Students should be encouraged to try a variety of learning strategies to find out what is most effective for them.

Learning Strategies

Metacognitive Strategies	Cognitive Strategies	Social/Affective Strategies
Students use as they plan for learning, monitor their learning, and evaluate their learning.	*Students use as they immerse themselves in the material to be learned.*	*Students use when they interact with others or with themselves.*

Metacognitive Strategies

Selective Attention
Student focuses on what the learning task will be and its purpose as set forth by the teacher, other students, or self.

Planning
Student makes plans for how learning will be accomplished—organizes materials, schedules time, and so forth.

Advance Organizers
Student reviews the material to be learned and identifies the major concepts. This may include pre-reading in a textbook or doing research.

Self-Monitoring
Student reviews the work in progress to make sure that comprehension is accurate, directions are being followed, and the final goal can be achieved.

Self-Evaluation
Student checks completed work to see if it fulfills the requirements of the assignment. If it does, what methods led to success? If it does not, what methods could be improved?

Cognitive Strategies

Rehearsal
Student reviews information needed to complete the task.

Transfer
Student identifies and applies prior knowledge to complete the task.

Translation
Student uses first language to aid in understanding the task.

Resourcing
Student finds resources (textbooks, dictionaries, encyclopedias, and so forth) that will provide information to help complete the task.

Grouping
Student classifies information based on its similarities and differences.

Recombination
Student breaks down larger concepts and recombines them to make them more meaningful or easier to understand.

Deduction
Student applies a set of rules to help in understanding and processing the material.

Note Taking
Student summarizes and records key concepts and key vocabulary.

Inferring
Student uses context to help determine meaning.

Imagery
Student uses images, either mental or on paper, to aid in understanding.

Social/Affective Strategies

Cooperation
Student works with other students to find solutions to problems, share information, or obtain feedback.

Question
Student asks teacher or other students for clarification or explanation.

Self-Talk
Student relies on self to provide assurance that the task can be accomplished successfully.

Self-Reinforcement
Student determines self-rewards to increase motivation for learning or completing a task successfully.

Developing Your Own Classroom Model

There are many approaches to teaching ELLs in the science classroom. Many teachers attribute their success to a combination of approaches, a healthy dose of trial and error, and an open line of communication with students. With this in mind, this section will provide a model and some tools for science teachers who have ELLs in their classrooms.

Organizing Your Approach

Before you begin your instructional plan, you will need to evaluate the proficiency of the ELLs in your science classroom. Your state, district, or school likely will have oral language test scores or evaluations available for each student. After you have gathered information on proficiency, you are ready to begin planning your lessons around specific content. Following are some steps you might use to develop your overall approach. It is likely that you are already incorporating many of these methods in your classroom. A model lesson follows this section.

Prepare

Select the major concept or theme of your lesson. Break down the lesson into manageable parts that will allow extra time for concepts that are more difficult. Analyze the material you will present, linguistically and cognitively, to identify areas that may be difficult to comprehend. Consider organizing the lesson on a concept map that can be shared with students.

Identify the scientific vocabulary that students will need to know. In addition to vocabulary words, consider other words that might become stumbling blocks such as *theory, calculate, interpret*, or *critical thinking*. Be prepared to point out some of the words in the textbook. Consider creating a vocabulary list or vocabulary cards that can be used by ELLs in the classroom.

Gather "props" that will help you convey meaning and process. The props can be used as you present the material or by students involved in independent study.

Gather and share pictures, diagrams, transparencies, and other printed materials that relate to the concept.

Gather the equipment and materials that students will need for activities and labs. Be prepared to model the products that students are expected to produce in their activity or lab work. You may want to fill out a sample lab report, a data table, or create a graphic organizer to show students what will be expected of them.

Motivate and Model

Determine prior knowledge. Prompt ELLs to talk about, draw, ask, or write questions, or convey in some other way what they know or think they know about this concept. Ask them to talk about or write about any personal experiences or personal analogies that might apply. Use props to help elicit responses.

Introduce the concept map you created during the ***Prepare*** stage to help you define and organize the information. Students can add information to it or fill in areas that you have purposely left blank.

Explain how the topic relates to the everyday lives of students.

Model what the student will be expected to know and walk them through the key points of the process or lesson. For example, if a laboratory experiment will be performed, you may want to review the type of equipment to be used and model the steps in the procedure. If students will be expected to produce a lab report, graph, diagram, or another product, model how they will be expected to do it. Ask questions to determine their level of understanding.

Give students a copy of a vocabulary list or vocabulary cards to use during the lesson or have them create their own device for learning vocabulary, such as a daily glossary or word journal.

Review with students the learning strategies that will be helpful in mastering the content of the lesson. Model or have other students model the strategies so students can begin to select what they think will work best for them.

Teach

After determining students' prior knowledge, you should have a general idea of what students know about the topic. This will allow you to structure your teaching around their knowledge.

Before presenting the lesson, review your oral presentation. Concentrate on presenting the material in a clear and logical sequence from beginning to end. Critique your choice of language to make sure that important words are not ambiguous or difficult to understand. Repeat or rephrase anything that may take time to comprehend. Make a mental note of areas that are most important and emphasize that content in your selection of tone, volume, and word choice. Look for visuals you can use to make the vocabulary and concepts comprehensible. For example, actual objects and pictures should accompany your lesson.

Clearly outline what students will need to know and be able to do after completing this lesson. Pass out a concept map, word list, or other written material that might assist students in taking notes or write the information on the chalkboard or on an overhead. When possible, refer to these aids to emphasize particular points and assist students in determining what is most important.

Summarize frequently and paraphrase as you go. Develop routines in your presentations that students will learn to recognize. When possible, call on students to do brief oral summaries or demonstrations in front of the class. Be sure to discuss and then correct any misconceptions that students might have about the topic. As you go, ask questions of students to elicit lower- as well as higher-order thinking skills.

It is important to provide information in all three formats: orally, in written form, and through visuals. Providing information in these three formats gives the ELL multiple opportunities to comprehend the lesson.

Provide time for questions or have students make a question notebook to share with you or an English-speaking student when time allows.

Practice

This is an essential stage for ELLs. It provides opportunities for them to practice what they have learned and allows them to reflect on what they do and do not understand. If you use scaffolding to present material to students, this will be a point where you can gradually pull back. Students can begin to make some decisions independently about their learning needs. This is also a good time to assign cooperative learning activities, pairing activities, and independent projects.

Provide as many hands-on activities as possible to ELLs. Research shows that ELLs exposed to a variety of activities make the most progress with their language skills.

Assess

All students, including ELLs, need to be accountable for what they learn in the classroom. Begin addressing assessment when you first introduce a new concept. ELLs need to know what is expected of them when it comes to homework, tests, reports, or other types of assessment.

Students should be given the opportunity to demonstrate what they have learned in many different ways—in writing and orally. Incorporate performance assessment tasks and other forms of authentic assessment when possible. It is helpful to use assessment tasks that are similar to those that were used to teach the material.

If possible, tailor assessment to individual needs. For beginning and intermediate students, it may be helpful to:

- ask students to response orally to questions
- give some open-book tests
- use the cloze procedure with graphic organizers
- use a matching format for tests or activities
- provide a word list to use with a fill-in-the blank activity

Always go over written and spoken directions to make sure they are understood. If needed, allow extra time for ELLs to do assessment tasks. Dictionaries might be helpful in some circumstances or word journals and lists. In some cases, ELLs may have to be tested on their science knowledge separately from their language knowledge. A self-evaluation log may be helpful for students to record their progress or track the areas in which they are having difficulty.

Evaluation

After all students have completed the lesson, it's time to evaluate what worked and what didn't work. Examine how students performed and determine if the methods you employed were successful. A professional log is helpful in tracking methods and can be used for future reference. Indicate what worked well and in which situations. This approach can help fine-tune your methods as you begin to see patterns emerge. Rely on your past evaluations to help you prepare for your next lesson.

Model Lesson: Simple Machines

Major scientific concept: motion and energy

Area of application: simple machines

Objective: After completing this lesson and the activities associated with it, students should be able to identify and describe the six types of simple machines.

Key vocabulary: lever, fulcrum, effort force, resistance force, pulley, wheel and axle, inclined plane, screw, and wedge

Science skills required: observing and inferring, classifying, forming operational definitions, comparing and contrasting, designing, formulating models, communicating, describing, making and using tables

Language skills required:

Reading: understanding technical vocabulary; understanding information on graphs, tables, and charts; understanding related material in textbook

Writing: note taking; labeling diagrams and charts; reporting on results

Listening: understanding explanations and technical vocabulary; understanding and participating in group activities; understanding directions

Speaking: answering questions; contributing orally to group activities

1 Prepare

- Bring in examples of simple machines or be ready to point out examples in your classroom: levers—bottle opener and capped bottle; nutcracker and nut; pulley—blind or curtain pull; wheel and axle—doorknob; inclined plane—wheelchair ramp; screw—screws; wedge—knife or chisel.
- Prepare a set of vocabulary cards to display in the classroom. Each card should include the vocabulary term and an illustration to accompany it. Be prepared to discuss definitions of words that might have dual meanings, such as *plane*, which might be confused with airplane or plain.
- Prepare the following concept map and make copies for student use throughout this lesson. Leave all or some portions blank so students can fill in as the lesson proceeds.

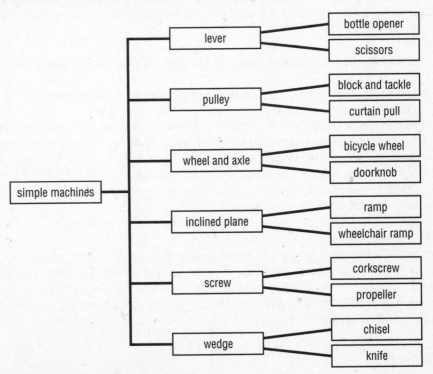

Gather materials for demonstrations and labs.

2 Motivate and Model

- Evaluate prior knowledge by showing students the examples of simple machines you brought to class. Provide stick-on labels for each of the machines. Ask students to write the name of the machine on the label or do it for them as they observe. Then have students group each type of machine by category—one pile for levers and one for inclined planes. Ask which machine should go into each category. (levers, pulleys, wheels and axles in one; inclined plane, wedge, and screw in the other)
- Show students the vocabulary cards you created. Ask for examples of other machines that fall into these categories. Explain that the purpose of machines is to make work easier.
- To illustrate how a machine makes it easier to move an object, help students with the following demonstration.
  1. Place a textbook near the edge of a table.
  2. Have students try to lift the book using only their fingertips.
  3. Then place the book on the end of the meterstick so that half of the book is resting on the stick, and the stick extends over the edge of the table.
  4. Have students push up on the meterstick to lift the books.
     Ask: *Which method required the most force to lift the books? The least?*
  5. Now point out and name the following:
     • the lever (the meterstick)
     • the fulcrum (the table edge)
     • the effort force (the hand)
     • the resistance force (the books)
     Next, say the name of each part and ask students to point to what you are saying. It is important that students recognize the new vocabulary before they are asked to produce the new words.
  6. Have students repeat the demonstration several times as they name the parts aloud.
- Tell students what they will be expected to do and learn in this lesson. Then
  • explain the use of the concept map (see *Prepare*)
  • explain and model the table activity (see *Practice*)
  • explain and model one or more steps involved in the pyramid activity (see *Practice*)
- Ask students to keep a log of questions that they have about the material. Make time to share the log with students or have English-speaking students collaborate to answer their questions.

3 Teach

- Before presenting the lesson, pass out copies of the concept map. Ask students to use the concept map as a guide for note taking.
- Review the key vocabulary and use transparencies or pictures to illustrate each type of machine. Rewrite the terms on the chalkboard as each is presented.
- Do a demonstration to illustrate the three types of levers — first, second, and third class. Show an example of each, then draw a diagram on the chalkboard or on an overhead transparency and ask students to label the position of the resistance force, the effort force, and the fulcrum on each.
- Conclude your lesson by asking a series of lower-order questions followed by a series of higher-order questions. Repeat and rephrase the answers given by students and draw a representation on the chalkboard with labels to reinforce the main points.

4 Practice

- Ask students to work in small groups to create a table that organizes information about the six kinds of simple machines —the type of machine, an example, and a brief description of how it works. If you did not model the table in the *Motivate and Model* stage, draw a sample table on the chalkboard. Give students a copy of a table rubric like the one that appears in the *Assess* section so they know on what standards they will be graded.

- Create a cooperative learning activity that allows students to use what they know about simple machines to make one or more hypotheses about how the Great Pyramids were built. Have students work together to create a model showing the use of one machine that might have been used to construct the pyramids. After the activity has been completed, have them check the validity of their conclusions by doing historical research. Give students a copy of a model rubric like the one that appears in the *Assess* section so they know on what standards they will be graded.

5 Assess

Use rubrics to help you assess student success on the table and model activities:

Rubric for Creating a Table

	Rating
1. The table includes the appropriate data.	
2. An appropriate title for the table is provided.	
3. The information in the table columns is appropriately organized and labeled.	
4. The information in the table is accurate.	

Rubric for Creating a Model

	Rating
1. A clear explanation is made of how the model will demonstrate the science concepts it is intended to show.	
2. A clear plan for the model is drawn. The plan shows dimensions and parts. Metric measurement is used.	
3. The plan includes an explanation of how the model simulates the real item. The explanation includes a description of how the model differs from the real item.	
4. The constructed model is sturdy and simulates the elements of the real item that it was intended to simulate.	
5. Color, labels, and other such devices clarify what the model is intended to show.	
6. The model is neat and presentable.	
7. The model is safe to use.	

- Ask students to evaluate what they have learned about simple machines and to identify areas where they need assistance by using self-evaluation checklists like the ones below.

What do I know about simple machines?

Six types:

1. _____ Example: _____

2. _____ Example: _____

3. _____ Example: _____

4. _____ Example: _____

5. _____ Example: _____

6. _____ Example: _____

Two categories of simple machines: _____

What science skills have I learned to use? What science skills do I still need help with?

	All of the Time	Most of the Time	Need more assistance
I am able to recognize and understand the vocabulary associated with simple machines.			
I am able to classify the different types of machines.			
I am able to describe the differences between the machines.			
I am able to observe how and why each machine works.			
I am able to develop a hypothesis about which machines were used to build the pyramids.			
I am able to interpret diagrams showing the simple machines.			

What language skills have I learned to use? What language skills do I still need help with?

Reading	All of the Time	Most of the Time	Need more assistance
I am able to read and understand the vocabulary words that describe simple machines.			
I am able to understand the information that appears on the simple machine concept map.			•
I am able to read directions.			
I am able to understand the information about simple machines in the textbook.			

Writing	All of the Time	Most of the Time	Need more assistance
I am able to take notes about simple machines that will help me study.			
I am able to label diagrams of simple machines with the appropriate words.			
I am able to create a chart or table that shows information about simple machines.			
I am able to write a hypothesis about the machines used to build the pyramids.			
I am able to describe and label a model. I am able to read and understand the vocabulary words that describe simple machines.			

Listening	All of the Time	Most of the Time	Need more assistance
I am able to understand most of what the teacher says about simple machines.			
I am able to recognize the vocabulary words when they are spoken.			
I am able to understand what other students say about simple machines in group activities.			
I am able to understand spoken directions.			

Speaking	All of the Time	Most of the Time	Need more assistance
I am able to discuss what I know about simple machines in front of the class.			
I am able to discuss what I know about simple machines in small groups.			
I am able to answer questions that the teacher asks about simple machines.			
I am able to describe what I have learned about the building of the pyramids.			

6 Evaluate

Evaluate your teaching strategies and identify areas in which you might want to change your approach to teaching about simple machines.

What methods or activities seemed to work the best?

_____

_____

What methods were not as effective? How could they be improved?

_____

_____

_____

What concepts seemed to be the easiest for students to comprehend? Which concepts were the most difficult?

_____

_____

What types of assessment were used?

_____

_____

Which type of assessment seemed most accessible to the students? Least accessible?

_____

_____

In the future, what types of assessment could be incorporated into this lesson?

_____

_____

Overall, how can I approve my approach the next time I teach this lesson?

_____

_____

_____

Teaching Resources

Many of the resources used with English-speaking students can be adapted for use with ELLs.

Graphic Organizers

Graphic organizers are valuable tools when working with ELLs because they present information visually. ELLs who have trouble grasping concepts by reading or listening can often use graphic organizers to increase their understanding. Graphic organizers can show patterns and relationships or help students organize important information. A description of several types of graphic organizers follows. Detailed instructions for making additional graphic organizers can be found in *Dinah Zike's Teaching Science with Foldables*.

Concept Maps

Concept maps have many uses for ELLs and can be used at several points in the learning process.

- They can be used to preview a topic or preview a chapter in a textbook. Students can visually record the concepts to be learned and organize them in a fashion that is most beneficial to them. Concept maps are especially useful when there are many new science terms for students to learn.
- They can be used as a framework for taking notes. Students can extend or revise their content maps to include new information and use them as a study tool.
- They can be used to review a topic. Constructing concept maps reinforces main ideas and clarifies their relationships.
- They can be used in assessment. Students can be asked to create a concept map or fill in a partially completed concept map as a way of assessing their knowledge of the material.

Below are samples of four concept maps that are useful in studying science.

Cycle Concept Map

Application:
- shows how a series of events interacts to produce a set of results again and again

CO_2 + H_2O

Sunlight/ Chlorophyll

Respiration — Energy released

Energy saved — Photosynthesis

O_2 + $C_6H_{12}O_6$

Events Chain

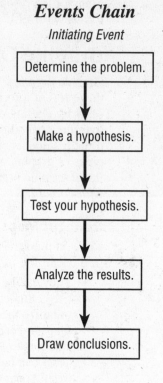

Applications:

- describes the stages of a process
- the steps in a linear procedure
- a sequence of events

Initiating Event

Determine the problem.

↓

Make a hypothesis.

↓

Test your hypothesis.

↓

Analyze the results.

↓

Draw conclusions.

Network Tree

Applications:

- shows causal information
- a hierarchy
- branching procedures

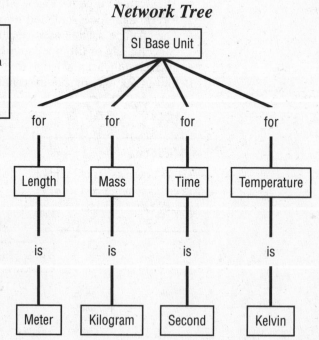

SI Base Unit

for — Length — is — Meter

for — Mass — is — Kilogram

for — Time — is — Second

for — Temperature — is — Kelvin

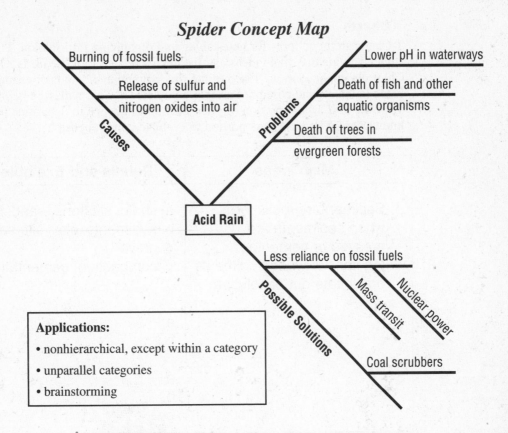

Spider Concept Map

Burning of fossil fuels

Release of sulfur and nitrogen oxides into air

Causes

Lower pH in waterways

Death of fish and other aquatic organisms

Death of trees in evergreen forests

Problems

Acid Rain

Less reliance on fossil fuels

Possible Solutions

Mass transit

Nuclear power

Coal scrubbers

Applications:

• nonhierarchical, except within a category

• unparallel categories

• brainstorming

K-W-L Charts

Creating K-W-L charts helps ELLs to take control of their learning. "What do I **KNOW**?" helps to establish prior knowledge of a topic. "What do I **WANT** to know?" arouses curiosity and helps the student set goals for learning. "What did I **LEARN**?" allows students to evaluate their own learning. These charts are effective in the *Motivate and Model* portion of the teaching sequence.

K-W-L Chart		
Topic: _____		
K **What I** **know about**	**W** **What I want to** **find out about**	**L** **What I learned** **and still need** **to learn**

T-Charts

T-Charts are good tools for note-taking and organizing information. They help students to identify main ideas and list details related to those ideas. On a T-Chart, the main ideas are listed in the first column and details or examples are listed in the second column. Partially completed T-Charts may be helpful in the *Motivate and Model* portion of your teaching sequence to determine prior knowledge or in the *Assess* portion to evaluate comprehension.

Main Ideas	Details and Examples
1. Sedimentary rocks form when sediments are pressed or cemented together or when sediments precipitate out of solution.	a. shale, siltstone, sandstone b. sediments: clay, silt, sand, gravel c. compaction, cementation
2.	a. b.
3	a. b. c.

Venn Diagrams

Venn diagrams can help ELLs compare the properties or characteristics of two items, two situations, or two concepts. The overlap areas not only help students to compare similarities and differences but also patterns and relationships.

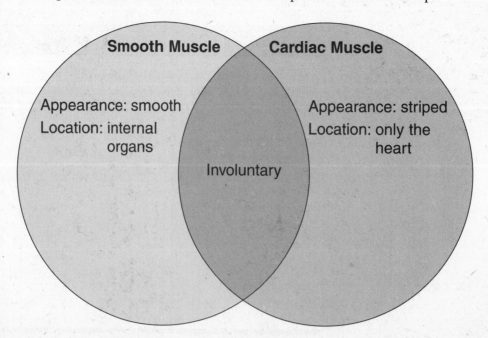

Smooth Muscle

Cardiac Muscle

Appearance: smooth

Location: internal organs

Appearance: striped

Location: only the heart

Involuntary

Other Classroom Tools

Cloze Procedure

The cloze procedure is a technique in which words are deleted from a passage. Students must fill in the words as they read. It is a good tool for assessing content knowledge, and it provides a model for proper sentence structure, word choice, and punctuation. When creating a cloze activity, select a passage that is self-contained and deals with a science topic that students are studying. It is usually best to leave at least the first and last sentence intact. Suggest that students read all of the passage first before filling in the blanks.

Crosswords

Crossword puzzles that incorporate key science terms provide good practice for ELLs. Students must learn definitions and incorporate proper spelling to be able to complete the puzzle.

Roundrobin and Roundtable

These techniques are useful in getting ELLs to participate in cooperative learning groups. Teachers give each group a topic and students take turns giving details about that topic. When done orally, it is called the roundrobin technique. The roundtable technique requires students to write their responses on a sheet of paper as it is passed around the group. Topics should be broad enough to allow students to respond three or four times in the circle. Some sample topics might include physical properties of matter, the periodic table, heredity, or pollution.

Learning Logs and Journals

Logs and journals are a good way to get ELLs to take the learning process into their own hands. They can provide teachers with information about how students are synthesizing and interpreting science topics. Teachers can review student logs and journals daily or collect them on a rotating basis.

Vocabulary Logs

By keeping a log of science terms and definitions, students can pinpoint words and concepts that are most difficult for them. They can then work independently to master them. Vocabulary logs do not have to be limited to key vocabulary words. They can be extended to include any word or language skill that requires more attention. Students can design their logs to fit their needs.

Question Journals

This type of journal provides an opportunity for teachers and students to "converse." Each day, or every few days, the student makes an entry about a science-related topic or issue. At first, ELLs may need to draw their entries instead of write them. The entry can convey a problem that the student is having with a particular concept or assignment, or it can ask specific questions. The teacher responds by writing in the student's journal. In addition to simply answering students' questions, teachers can provide examples or study tips, or even anecdotes to engage the student's interest. If students do all or some of their journal writing in their first language, teachers can get help with translation from other ELLs, parents, or ESL teachers. As students get more comfortable with writing in English, teachers can encourage them to change to writing in all English. This process is a good way to monitor a student's progress with written language skills.

Bibliography

Abdi, S. Wali. (1997). "Multicultural Teaching Tips," *The Science Teacher.* Feb. 1997.

Alexakos, K. (2001). "Inclusive Classrooms," *The Science Teacher.* March 2001.

Anstrom, K. (1998). "Preparing Secondary Education Teachers to Work with English Language Learners: SCIENCE," *NCBE Resource Collection Series,* No. 11, Dec. 1998. Washington, DC: National Clearinghouse for Bilingual Education (NCBE).

Behm, C. (2001). "Big Picture Science: Uncovering teaching strategies for underrepresented groups," *The Science Teacher.* March 2001.

Blakey, E. and Spence, S. (1990). "Developing Metagcognition." ERIC Digest. Syracuse, NY: ERIC Clearinghouse on Information Resources.

Buchanan, K. (2001). *School Administrator's Guide to the ESL Standards.* Alexandria, VA: Teachers of English to Speakers of Other Languages (TESOL).

Carlson, C. (2000). "Science Literacy for All," *The Science Teacher.* March 2000.

Chamot, A.U., and O'Malley, J.M. (1994). *The CALLA Handbook: Implementing the Cognitive Academic Language Learning Approach.* Reading, MA: Addison-Wesley.

Crandall, J. (1994). "Content-centered Language Learning." Washington, DC: ERIC Clearinghouse on Languages and Linguistics.

Curtain, H. and Haas, M. (1995). "Integrating Foreign Language and Content Instruction in Grades K-8." Washington, DC: ERIC Clearinghouse on Languages and Linguistics.

ESCORT. (2001). *The Help! Kit: A Resource Guide for Secondary Teachers of Migrant English Language Learners.* State University College at Oneonta, NY.

Hargett, G. (1998). "Assessment in ESL & Bilingual Education." Portland, OR: Northwest Regional Educational Laboratory's Comprehensive Center, Region X.

Laturnau, J. (2001). "Standards-Based Instruction for English Language Learners." June 2001. Honolulu, HI: Pacific Resources for Education and Learning.

Lessard-Clouston, M. (1997). "Language Learning Strategies: An Overview for L2 Teachers." Nishinomiya, Japan: Kwansei Gakuin University.

Lindholm-Leary, K. (2000). *Biliteracy for a Global Society: An Idea Book on Dual Language Education.* Washington, DC: National Clearinghouse for Bilingual Education (NCBE); Center for the Study of Language and Education; Institute for Education Policy Studies; Graduate School of Education and Human Development; The George Washington University.

National Science Education Standards. (2000). Washington, DC: National Academy Press.

NCBE. (2000). "Teaching Science in a Dual Language Classroom." Washington, DC: National Clearinghouse for Bilingual Education (NCBE).

Reyhner, J, and Davison, D. (1993). "Improving Mathematics and Science Instruction for LEP Middle and High School Students Through Language Activities." Washington, DC: National Clearinghouse for Bilingual Education (NCBE).

Rivera, C. and Stansfield, C. (2001). "The Effects of Linguistic Simplification of Science Test Items on Performance of Limited English Proficient and Monolingual English-Speaking Students." Paper presented at the Annual Meeting of the American Educational Research Association, Seattle, WA, April 12, 2001.

Simich-Dudgeon, C. and Egbert, J. (2000). "Science as a Second Language: Verbal interactive strategies help English language learners develop academic vocabulary," *The Science Teacher.* March 2000.

Spangenberg-Urbschat, K. and Pritchard, R. eds. (1994). *Kids Come in All Languages: Reading Instruction for ESL Students.* Newark, DE: International Reading Association.

Stansfield, C. (1992). "ACTFL Speaking Proficiency Guidelines." Washington, DC: Center for Applied Linguistics.

Sutman, F., Allen, V., Shoemaker, F. (1986). *Learning English Through Science: A Guide to Collaboration for Science Teachers, English Teachers, and Teachers of English as a Second Language.* Washington, DC: National Science Teachers Association.

Tannenbaum, J. (1996). "Practical Ideas on Alternative Assessment for ESL Students." Washington, DC: ERIC Clearinghouse on Languages and Linguistics.

Tomlinson, C. (2001). *How to Differentiate Instruction in Mixed-Ability Classrooms,* 2nd Edition. Alexandria, VA: Association for Supervision and Curriculum Development (ASCD).

Walker, B., Scherry, R., and Gransbery, C. (2000). *Collaboration for Diverse Learners: Viewpoints and Practices.* Newark, DE: International Reading Association.

Williams, J. (2001). "Classroom conversations: Opportunities to learn for ESL students in mainstream classrooms," *The Reading Teacher.* May 2001.